ENAMEL
without
HEAT

**By Stephen J. Schilt
and Donna J. Weir**

STERLING
PUBLISHING CO., Inc.   New York
SAUNDERS OF TORONTO, Ltd., Don Mills, Canada

Oak Tree Press Co., Ltd.
London & Sydney

739.15

Little Craft Book Series

Copyright © 1971 by Sterling Publishing Co., Inc.
419 Park Avenue South, New York, N.Y. 10016
British edition published by Oak Tree Press Co., Ltd., Nassau, Bahamas
Distributed in Australia by Oak Tree Press Co., Ltd.,
P.O. Box 34, Brickfield Hill, Sydney 2000, N.S.W.
Distributed in the United Kingdom and elsewhere in the British Commonwealth
by Ward Lock Ltd., 116 Baker Street, London W 1
Originally published © 1971 by Saddleback Publications,
Santa Ana, California
Manufactured in the United States of America
All rights reserved
Library of Congress Catalog Card No.: 75 167658
ISBN 0-8069-5180 X   UK 7061 2323 9
5181 8

Enamels + enameling

A sunrise on glass—colored not with chemicals, as stained glass is, but with odorless, easy-to-apply heatless enamel.

Contents

Illus. 1. A panel from the cathedral at Chartres? Actually, this is a "stained" glass panel that you can make yourself, without heat, chemicals, or other elaborate equipment.

Before You Begin

Smooth, shiny and colorful, an enamel glaze is one of the most beautiful finishes you can apply to a plain surface. And an enamel glaze is practical as well: scratch-resistant and heat-resistant, enamel protects as it decorates jewelry, trays, or pottery.

Not many craftsmen who work at home become involved with enamelling, because of the need for an expensive kiln in which to bake the enamel powder. The co-author of this book, Stephen J. Schilt, worked for a number of years to invent something better than enamel for craftsmen to use. He developed a product that, when mixed in the proper proportions and allowed to harden, looks, feels and acts just like enamel— but you apply it without heat! Called "Boss Gloss™," the heatless enamel adheres to any surface, including those which cannot be enamelled by baking. Wood, clay, metal, glass, hard plastic— and even plastic foam—are only a few examples of the surfaces you can decorate with this one-step substance. Even children can safely apply an enamel-like finish to any surface with this new, non-toxic enamel. The discovery of this new innovative product is bound to have a profound effect on many crafts, and this book is being published to acquaint you with the method of enamelling without heat.

Mr. Schilt's company, California Titan Products, Inc., Santa Ana, California, manufactures this heatless enamel, and it is found in most art and craft supply shops. The heatless enamel comes in two containers. All you do is mix the two liquids together, brush or drip the mixture on any surface, and set the decorated object aside in a warm dry room until the enamel hardens. This usually takes a few days, the exact amount of time depending upon how thick a layer of enamel you want to apply. When the surface is hard the heatless enamel is solid and non-tacky as any baked glaze. The finish does not melt or scratch; it is water- and alcohol-proof; and it needs only occasional dusting to keep its surface shiny and attractive.

By mixing the heatless enamel yourself, you control the density of color and thickness of the medium. In one jar is the POLYMER (numbered No. 1 on the jar), that part which gives the enamel its color. It is used in its liquid state. A wide variety of colors is available, including white, and by mixing the colors you can produce any unusual shade you require.

Illus. 2. A jar of Polymer, the component of heatless enamel which gives color to the mixture.

5

Illus. 3. Clear Extender is a thinner, colorless form of Polymer. Use it in place of Polymer, or mix it with Polymer for a lighter shade.

Illus. 4. Without Curing Agent, the heatless enamel will not harden. Mix 1 part Curing Agent with 3 parts Polymer or Clear Extender for a permanent surface.

CLEAR EXTENDER (No. 3) is a variation of Polymer (No. 1). It is thinner than Polymer and has no color. Use Clear Extender to thin the Polymer for easier flow and to lighten the color. You can also use Clear Extender by itself with Curing Agent (see below) to make a clear coating or an overglaze.

CURING AGENT (No. 2) is the substance which combines with the Polymer or Clear Extender so the surface hardens. Also a liquid, Curing Agent must always be combined with Polymer and/or Clear Extender in the ratio of 1 to 3—that is, 1 part Curing Agent to every 3 parts Polymer or mixture of Polymer and Clear Extender or Clear Extender alone. *You must follow these proportions or the heatless enamel will not dry and harden properly.*

The beautiful colors and glossy finish of heatless enamel resemble baked enamel glazes so closely that even experienced craftsmen are fooled. And you can decorate many objects—trays, dishes, jewelry, tiles, lampshades, glasses, table tops. Found in most craft and hobby shops, heatless enamel even coats surfaces which cannot be heated in a kiln. When you design with heatless enamel, you are not only creating beautiful ornaments, but enjoying hours of fun and colorful excitement as well.

Decorating Plastic Foam

Illus. 5. Decorate plastic foam with heatless enamel for an animal, storybook character or bizarre creature from outer space.

Plastic foam, commonly known in the United States by its trade name, Styrofoam, has challenged craftsmen who tried to decorate its surface. Most paints contain solvents which corrode the delicate foam, however, so that up until now only water paints have been used. Unfortunately, the finish of water paints is very flat, and the paint can only be applied in thin layers. Heatless enamel is a tremendous boon to would-be plastic foam decorators, since it contains no solvents which attack the foam, has a glossy finish, and can be applied in any thickness.

Buy some plastic foam in a craft shop or variety store. It is inexpensive and easy to cut with a knife or razor. Lightly sketch an outline on the plastic foam with a pencil, taking care not to press too hard. Cut around this outline with a knife in a holder or a single-edge razor blade. Plan your colors and, if you want a special shade, mix several colors of Polymer to make new tones. Place the plastic foam figure on a flat surface—the piece must be perfectly level or the glaze will not dry smooth.

If you do not want the bubbly surface of the plastic foam to show through colored glaze, coat the foam with Clear Extender or Gesso. Pour 1 ounce of Clear Extender into a mixing cup (use either disposable paper or plastic cups, as the mixture sticks to all surfaces) and then add 2 teaspoons of Curing Agent. There are 6 teaspoons per ounce, so you are mixing Extender and Curing Agent in the proper ratio of 3 to 1. Mix the Clear Extender and Curing Agent together by stirring thoroughly with a stick or plastic spoon, but avoid stirring air into the mixture. Allow the mixture to sit for at least 5 minutes, so that the two substances have a chance to combine completely. Use a paint brush to apply the clear mixture.

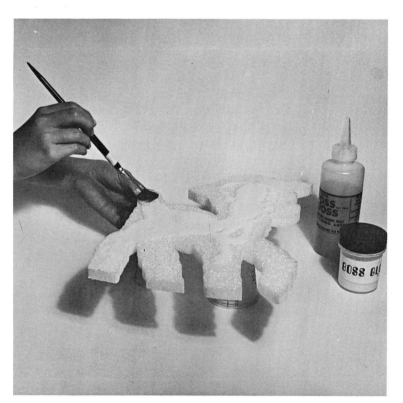

Illus. 6. To smooth the textured surface of the plastic foam, brush on a layer of Clear Extender mixed with Curing Agent and let dry for a day.

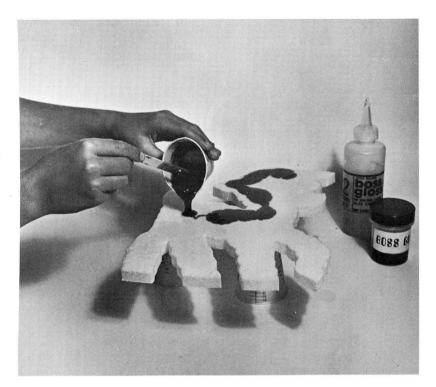

Illus. 7. When the layer of Clear Extender is dry, pour a mixture of colored Polymer and Curing Agent on the surface.

Allow this clear layer to dry for about a day. Heatless enamel dries according to schedule every time, so you can expect the following stages in the drying process:

First is the induction stage, which occurs right after mixing Polymer or Clear Extender with Curing Agent. The mixture is the same consistency as it was in the jar, either liquid or gel. After 1 to 3 hours, the compound becomes a thick, syrupy liquid. After 5 to 7 hours, the enamel is the consistency of taffy and you can no longer apply it with a brush. Instead, use a spatula or palette knife. In the gel stage, after 8 to 10 hours, the coating appears wet but it is gelled. If you touch the glaze now it stays where you push it. In the semi-cured stage, after 20 to 24 hours, the surface is fairly hard and can be gently handled. Complete curing of the surface takes about 7 days, but it is worth the wait: the surface is hard, scratch-resistant and permanent.

If you apply a first coat and want it to dry before you apply the second, leave it for a while—

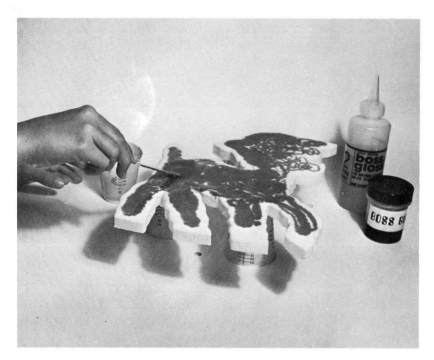

Illus. 8. Push the Polymer to the edges of the piece with a stir stick. It is too thick to use easily with a brush.

preferably for 20 to 24 hours, but for at least 8 hours. Be careful when you finally touch it that you do not handle it too strongly or the layer will move.

For the colored layer, mix colored Polymer with Curing Agent in the same proportions you mixed Clear Extender—3 to 1. Pour or brush the mixture on the plastic foam shape and set the piece aside to harden. After a day, apply facial features (if your shape is an animal or a person) or other small details (if you have cut out an inanimate object). Use either a paint brush or a toothpick for fine areas. Leave the plastic foam piece in a warm dry room for about a week to allow it time to harden completely.

NOTE: If you leave a project to harden in a humid area (a steamy kitchen or a damp garage, for example), the surface of the glaze may be dulled with a "blush." To make sure the glaze dries to an attractive and clear finish, always place your projects in a dry room to harden. If a blush does form, apply a layer of Clear Extender to restore some of the shine. The true color, however, might not be restored.

For variety, outline the plastic foam and add details to the surface with narrow cord or braid.

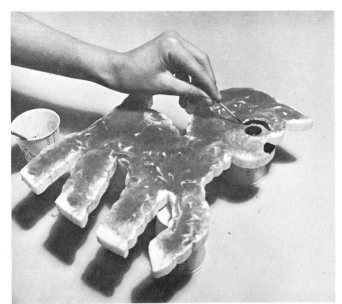

Illus. 9. After the Polymer has slightly hardened, add small details (facial features and hoofs are added here) in another color with a toothpick.

Illus. 10. To define areas of your plastic foam fellow, outline them with dark cord or braid.

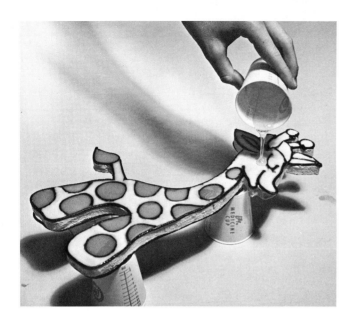

Illus. 11. For a level surface, pour Clear Extender mixed with Curing Agent on the colored areas until they are even with the raised braid.

First apply the clear base coat and then, using white glue and a toothpick, attach the braid around the outside of the figure and to the face. Turn to the perky giraffe in color on page 26 to see how dark braid enlivens a simple figure.

After attaching the braid, fill in each area with glaze and let the piece dry overnight. For additional gloss, and to bring the surface of the glaze level with that of the braid, add an overglaze of Clear Extender all over the plastic foam.

At a loss as to what to do with your decorated plastic foam? Use small pieces as coasters—moisture does not soak through hardened heatless enamel—or a larger piece as a decorator tray for lightweight items. Make simple toys from plastic foam just as you would with wood: cut figures and add heatless enamel features with a brush.

Glue the figures to sticks, for stick puppets, or simply let the heatless enamel harden and play with two-dimensional dolls. If you want to make a large project, construct a doll house for your plastic foam family. You do not need nails to fasten plastic foam, only glue—and instead of flat, dull paint, use shiny Boss Gloss for shingles on the roof, windows, doors, and even shrubbery around the base. Cut out seasonal shapes and add appropriate colors: a plastic foam Christmas tree, with a background of green heatless enamel and small red drops as festive ornaments, makes a lively addition to your holiday decorations. Make an Easter bunny, a Halloween pumpkin, a Valentine's Day heart. A plastic foam decoration lasts longer than paper, and the heatless enamel surface looks and acts better than paint!

Jewelry

Applying heatless enamel to jewelry findings—those undecorated metal shapes which you can buy in craft and hobby shops—is even easier than applying it to plastic foam: the metal findings are already smooth and need no base coat under the colored Polymer. You are of course not limited to jewelry; you can use these decorating suggestions on metal trays, coasters and dishes too. And the type of metal you use does not matter. Use brass, steel, aluminum foil, copper, tin or any alloy—just be sure that the surface is clean and free of grease before you add the heatless enamel. To clean the metal, soak it in 1 cup vinegar with 1½ tablespoons salt for a few minutes.

Be sure the piece you are working on is perfectly horizontal. If there is already a clasp or loop on the back of the finding, press it into a piece of plastic foam or lay the finding on an inverted paper cup so the surface of the finding is level. Mix the Polymer and Curing Agent in the proportions of 3 to 1, as instructed on page 8. Thin the Polymer with Clear Extender if you want, to help the mixture flow easily on to the small surface of the finding.

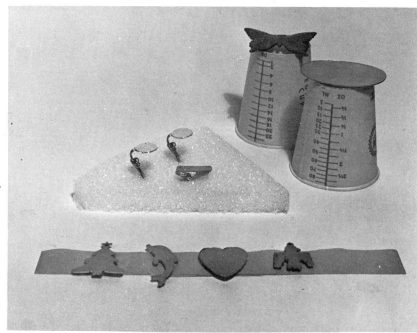

Ilus. 12. Place jewelry findings on a level surface. The small shapes in the foreground lie on double-sided sticky tape to keep them steady.

Illus. 13. Clean the surface of the metal finding. Then drip some heatless enamel mixture and push it to the borders of the shape.

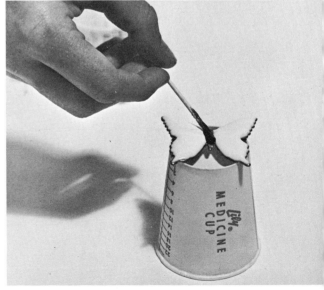

Illus. 14. Fill in small areas with a toothpick. Though you can add details while the main portion is still wet, you risk smudging the colors.

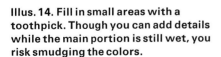

Because the area to be covered is small, it is easier to drop the heatless enamel on the finding with a stir stick or toothpick than to brush it on. Gently push the mixture to the edges of the jewelry piece. If bubbles appear during the first few hours of drying, as they might if you mixed your solution too vigorously, puncture them with a toothpick. Because the heatless enamel has not hardened yet, the spot where the bubble was fills in by itself.

To blend colors together, drop more than one color of Polymer/Curing Agent mixture on the jewelry finding. Swirl the colors together with a toothpick or leave them alone to keep them separate. To make thin strings of color that stretch across the surface of the finding, mix some enamel in a cup and let it reach the syrupy stage (1 to 3 hours). Dip a toothpick into the cup and quickly pull it out. Lay the string that forms when you remove the toothpick across the

Illus. 15. Lay a string of the syrupy mixture over a still-wet surface. When everything is dry, the two colors are fused together.

surface of the finding, already covered with a layer of enamel. As you pull the toothpick from the cup, the string that comes with it should be firm and evenly shaped. If the heatless enamel is still too runny, wait a while—perhaps an hour—and try again.

A less risky way to add strings of color is to form them separately before you apply them to the finding. Dribble the heatless enamel mixture on freezer paper or waxed paper in thin strings or round drops. Leave these shapes to cure (at least 20 to 24 hours), and then gently peel the paper from them. If the heatless enamel has not completely hardened, you can cut and trim the shapes. Place these pieces in just the right position in the wet enamel on the finding, and let the entire

Illus. 16. Swirl colors and patterns with heatless enamel as you would swirl icing on a cake. Just one stroke creates interesting patterns.

Illus. 17. Preformed shapes—drops and strings that you let harden on waxed paper—lessen the possibility of mistakes. Place them on wet enamel and let harden.

piece dry completely. When the glaze has hardened, attach pins, rings or loops to the finding for a piece of jewelry.

If you are decorating several findings to use on a bracelet or necklace, apply heatless enamel to their surfaces *before* you attach them to a chain. This reduces the risk of mistakes and gives you greater freedom to turn the findings around as you decorate them.

Glaze on Clay and Natural Surfaces

Heatless enamel is so durable and non-porous that when you apply it to greenware (unbaked clay), you eliminate the need for baking. The glaze completely covers the surface and seals in moisture so the clay does not dry out and crack. Of course, a ceramic-like heatless enamel glaze is perfectly suitable—and extremely beautiful— on bisqueware (baked clay), as well as on plaster of Paris, modelling clay and even papier mâché.

To decorate greenware, carefully trim and smooth the piece as if you were going to fire it in a kiln. Once you apply the heatless enamel, there is no chance to reshape the clay under it, so be sure your piece is as smooth as can be.

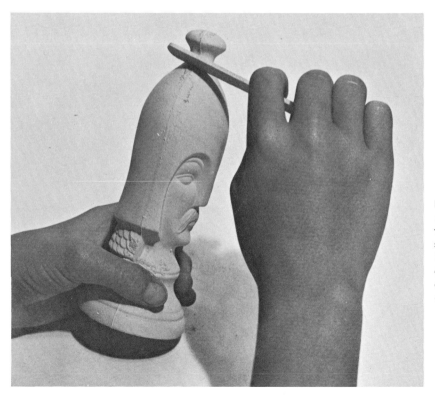

Illus. 18. Even if you have no ability in modelling, the unusual abstract shapes you make can benefit from a coat of color. Smooth the piece completely before you even mix the enamel.

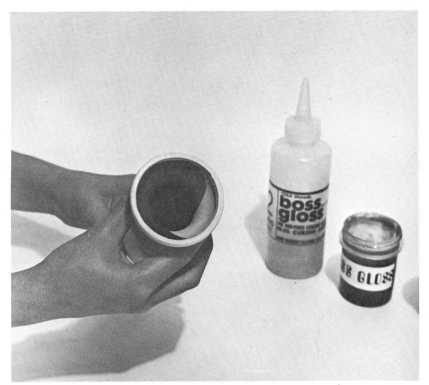

Illus. 19. To coat the inside of a hollow form, turn the piece around and around. Keep turning as you pour the excess into a paper cup. If the last few drops make bubbles on the rim, pop them with your finger or a toothpick.

If the pottery is hollow—perhaps a bowl, vase or a figurine with a hollow base—glaze the inside first. Mix Polymer and Curing Agent together in the proportions of 3 to 1, and pour a small amount of the mixture into the hollow portion. Swirl the mixture around so the inside becomes coated with heatless enamel, and pour any extra out slowly. Heatless enamel does not keep, so either use it right away on another project, or discard it.

Allow the inside of your pottery to cure overnight, and then apply heatless enamel to the outside of the piece with a brush. Keep your brush strokes vertical and even. Use a liberal amount, but not so much that the mixture runs. You can always add a second coat later; in fact, on greenware and air-dried clay, a second coat is a good idea. Cover every portion except the bottom of the pottery with heatless enamel. Let the glaze harden for several days—the longer the better—and gently place the piece upside down in a large cup. Coat the bottom with heatless enamel and leave the pottery to harden completely.

Add other colors with previously formed bits of heatless enamel to the pottery as you did to the jewelry findings (page 16). Shaped pottery usually does not need too much decoration other than a layer of glaze, since its beauty comes from the shape of the piece and the gloss of its surface, not in fancy colors and designs. On flat shapes of air-dried clay, however, designs on the surface add much to the charm of the piece. You might feel you are icing a small cake, instead of applying a permanent, non-porous finish to clay!

Other natural surfaces gain much appeal when you color them with heatless enamel. Add a face or interesting design to a smooth flat stone and use the finished piece as a paperweight, doorstop, or coffee table knickknack. Sea shells often have interesting shapes and need only a few highlights

Illus. 20. Because the pottery is hard and heavy, it supports the pressure necessary to brush the glaze on. Use long, even strokes, and be careful to reach all corners of the surface.

Illus. 21. Brush heatless enamel on a small clay or stone shape to use as a paperweight, door-stop or other ornament. If you get some enamel on your hands or the table while you work, wash it right off before it hardens.

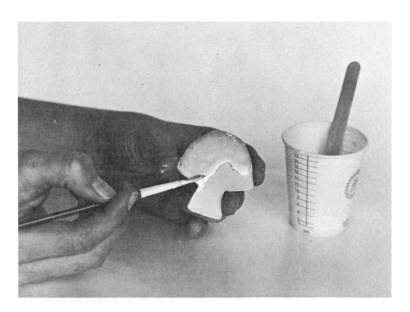

Ilius. 22. Decorate insects, birds and trees with preformed enamel drops for exact yet casual placement of color.

of color to emphasize a face or pattern in their texture. If you apply heatless enamel to wood, be sure the piece is clean of all sand and dust, and that it is smooth. Rub it with fine-grained sandpaper to make sure the surface is even. The same rules for mixing and drying hold when you decorate these natural surfaces with heatless enamel. Any surface you can paint, you can enamel without heat!

Illus. 23. Is it a gingerbread man? No, it's a little clay man dressed up in a heatless enamel costume.

Decorating Tiles

Illus. 24. Making a cheeseboard of tiles sounds much more difficult than it really is. Color each tile separately —and simply—with heatless enamel, and then cement them together.

Heatless enamel transforms plain tiles into colorful objects which are, like everything covered with the enamel, durable, decorative and practical. You can buy a ceramic tile which has not been glazed (in bisque state) and decorate it with an enamel design, or you can make your own tile from clay and cover its surface with heatless enamel. Use the individual tiles as trivets or coasters (heatless enamel resists heat and moisture) or glue several together and set them in a base for a cheeseboard or tray. A simple yet glamorous cheeseboard is shown above and in color on page 27.

Cheeseboard

The colorful cheeseboard on page 27 uses 12 tiles, each $4\frac{1}{4}''$ square, which were purchased already fired but not glazed. The colorful design was planned so the colors would blend from red to yellow, and from blue to yellow. Decorate each tile separately, but be careful that when the tiles are placed next to each other, they fit together exactly. Consider each tile like a part of a mosaic or a patchwork quilt. You might even plan a picture, instead of an abstract design, but be sure to position every part of the picture on each tile in the exact position it belongs.

Use any decorating tricks for the tiles that you have learned so far: pour strips of one color on the tile, and add another color while the first is still wet. Make the colors bleed together by drawing one into the other with a toothpick. Or, to keep the colors distinct, allow the first strip to gel slightly before you pour the next.

Take advantage of the semi-cured stage of heatless enamel to add new designs. When the glaze is only slightly tacky, cover the surface with a textured vinyl paper. Press the vinyl firmly into the glaze so its raised pattern transfers to the surface of the tile. Leave the vinyl on the tile until the heatless enamel has completely hardened, and then carefully peel the vinyl from the tile. The surface of the tile is marked with the regular relief pattern of the vinyl! This textured surface is unobtainable with any other type of glaze.

After the decorated tiles are dry, make the wooden base to hold them. Buy a piece of plywood which is the size of all the tiles put together, in the shape you want the tray to be. Cover the

Illus. 25. The tile is supported by two inverted paper cups to keep it level and away from the table. Pour or brush heatless enamel on the smooth surface.

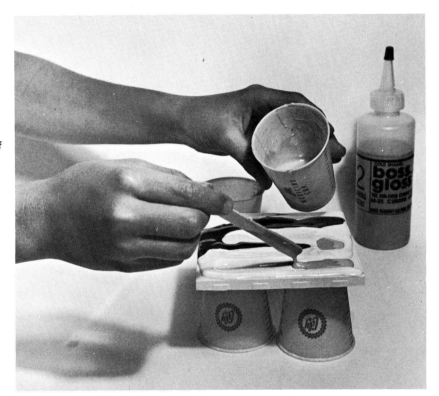

Illus. 26. Use a stir stick to drop small amounts of enamel on the tile, and pull the enamel to the edges.

plywood base with one or two layers of shellac, to waterproof the wood. Place the tiles on the plywood, but do not glue them in place yet. You might want to surround the tiles with wooden strips first. The cheeseboard on page 27 is framed with walnut strips, which give the board a rich appearance. Use any wood which looks attractive to you. Shellac the strips for extra sheen if you want, and glue them to the plywood base with a strong, waterproof glue.

To attach the tiles to the tray permanently, use a cement made especially for clay tiles (your craft shop can supply you with one). Apply cement to both the plywood base and the back of the tile, and place the tile on the base. If there are any *interstices*—small spaces between the tiles—fill them in with cement. After the cement has dried, you can wash your cheeseboard, cut on it, and handle it as roughly as you would any store-bought tray. This one, however, is your own design and your own creation!

Illus. 27. Bright colors and lively patterns make an extraordinary menagerie of plastic foam animals.

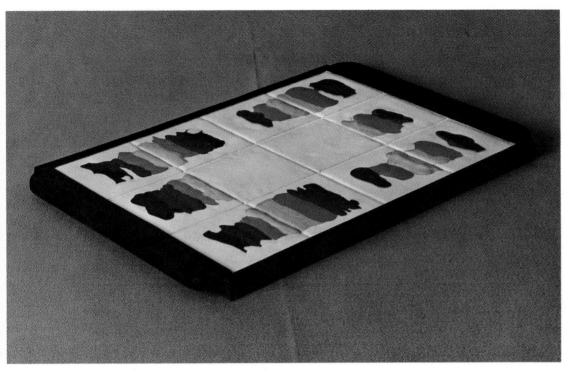

Illus. 28. The abstract design on this cheeseboard does have a regular pattern: the colors blend from red to yellow to red on one half, and from blue to yellow to blue on the other.

Cloisonné Technique

Not only does the cured surface of heatless enamel resemble baked enamel, but the decorating techniques which are used for the two glazes are similar. *Cloisonné,* up until now used only in enamelling, uses preshaped wires to separate areas of color. You can achieve sharp color contrasts without having to wait for the first color to gel before you apply the second. The cloisonné technique is very popular in jewelry-making, but try the technique on tiles first: they are larger and easier to work on.

Brush a layer of heatless enamel on a bisque tile. While it starts to gel, cut and bend thin steel wire into different shapes. Use a pliers to bend

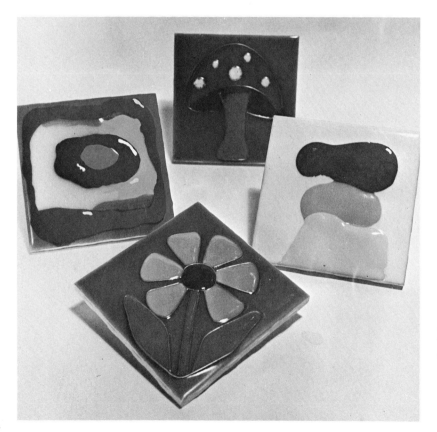

Illus. 29. Colors applied either at random or carefully separated by fine wires both have a decorative charm, but the effect of each is quite different.

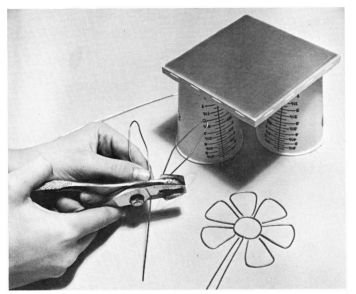

Illus. 30. Carefully cut and bend the wire into the proper shape. Be sure each section lies flat.

Illus. 31. Place each piece of wire on the enamel surface while it is still slightly tacky.

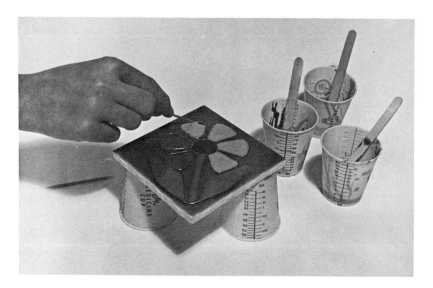

Illus. 32. When the base coat of heatless enamel is dry, fill in the sections between the wires with other colors. Spread the enamel to the edges of each section with a toothpick.

and a wire cutter to cut the wire. *Be sure that all the pieces of wire lie flat.* Once you set them in the enamel, you cannot change their curves.

About 3 hours after applying the base coat, lay the wire pieces in place. When the base coat is semi-cured, fill in each wire "cell" with colored Polymer and Curing Agent (proportions of 3 to 1). Push the Polymer to the edges of the wire sections with a toothpick or small brush, and allow the tiles to harden undisturbed for one week. Use these individual tiles as trivets, wall decorations, or parts of a larger tray.

Illus. 33. Glossy or dull, transparent or opaque, abstract or realistic—heatless enamel lets you design any way that pleases you!

"Stained" Glass Panel

Elegant stained glass windows have decorated churches and rich-looking homes for centuries. The color changes that occur when sunlight shines through stained glass are not equalled in any other medium—but transparent heatless enamel on glass comes very close to the real thing! And a window decorated with heatless enamel can provide the ornamental touch that changes a plain "furnished" room into a room that is "decorated."

Cut a sheet of glass to the proper size, or have a professional glass cutter do this for you. Before you begin mixing colors, plan a design for your panel on paper. Do not make the design too detailed, since even though heatless enamel is liquid, it does not flow as easily as paint. Place the paper design under the glass panel.

To make the glass panel look even more like

Illus. 34. Drill holes in the glass panel and suspend it from the ceiling, or mount it in a window or on a table. Wherever you place it, the glass panel needs sunlight to show off the colors best.

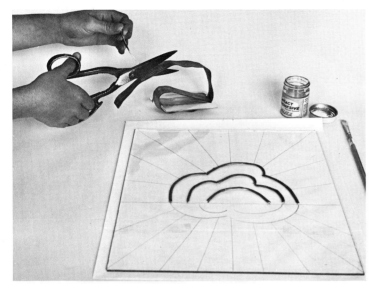

Illus. 35. Cut the lead came with sturdy scissors. Lead is a soft metal, so you can twist and curve it to fit your lines exactly.

real stained glass, cut and glue lead strips (called *came,* found in craft shops) along the lines of your pattern. The lead you buy will probably be wider than the line you want to divide the colors; use a pair of sturdy scissors to cut the lead to a narrower width. Attach the strips with lead adhesive to the glass above the lines of your pattern.

Illus. 36. Place the paper diagram under the glass and glue the lead strips along the pattern lines.

Illus. 37. Jewelry decorated with heatless enamel spills from a treasure chest. Perhaps the wooden lid is also colored with enamel!

Illus. 39. A small tray, unassembled tiles, and a portion of the glass panel from page 32 are the ornaments shown here. The glossy surface of these objects vibrates with color.

Illus. 38 (left). The greenware on page 18 has been smoothed, brushed with heatless enamel, and allowed to dry. The result is this somber but shiny face.

Illus. 40. Remember to measure the Polymer and Curing Agent carefully. While the mixture should be kept only in disposable containers, you can use metal or glass for measuring, since a single component does not harden.

Illus. 41. After the lead adhesive has dried, mix the heatless enamel and drop it into the proper sections of the glass. Since the lead separates the colors, you do not need to wait for each area to dry before applying the next.

Illus. 42. In areas where two colors meet, play safe by letting one gel slightly before applying the next.

When the adhesive has dried (see the package instructions for the approximate time), mix the enamel Polymer and Curing Agent together in the proportion of 3 to 1 and apply the different colors to the glass. The lead strips separate the different colors just as the braid did on your plastic foam figure (page 10), so you can apply all the colors at one time without waiting for each one to gel. Depending on how detailed your pattern is, use either a brush or a stick to apply the glaze. Set the glass aside to cure.

The sunlight shining on your panel shows off your craft best—particularly if you use transparent colors. Light opaque colors also create interesting effects when used in the middle of a transparent pane, and can also simulate "milk glass." For another effect, mix a transparent Polymer with a small amount of opaque color to produce a smoky, translucent panel.

Personalize drinking glasses with names in heatless enamel so you can tell whose drink is whose. Color small pieces of glass, drill a hole in each, and hang them like a mobile. The sound they make as they gently bump each other is melodious, and the sight they offer in the sunlight is colorful and cheery.

Illus. 43. Try your hand at forming flowers: paint a thin layer of heatless enamel on waxed paper in the shape of a flower. When the enamel is dry, peel the paper off.

Illus. 44. Three little men who look like iced cakes are really clay figures coated with heatless enamel!

Lamp Shade

Tiffany lamp shades were most popular at the end of the nineteenth century, and lately there has been a revival of interest in the delicacy of their jewel tones. As authentic Tiffany shades are quite costly, why not design and decorate your own shade to look like the real thing? Heatless

Illus. 45. The room gains a new mood when the light shines through this lamp shade.

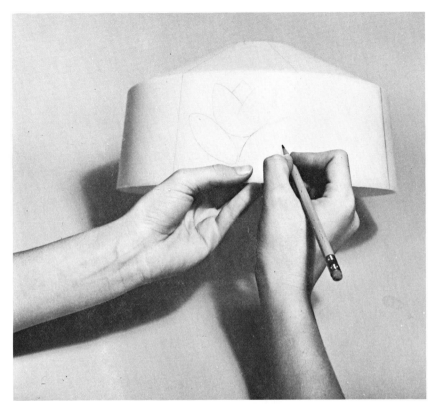

Illus. 46. The lamp shade which is being pencilled here comes in two parts: the bottom shade and the top crown, which hides the ugly fixture. Draw and decorate each part separately.

enamel on a plastic shade dries to a smooth, shiny finish, and the artificial light from underneath reflects off the glazed surface to highlight the shining colors.

Your craft shop has several styles of plastic shades ready for decorating, including the traditional Tiffany shape, or you can buy a cylinder or hemisphere directly from a plastics' distributor.

If the surface of the plastic is very smooth, roughen it slightly with sandpaper so the heatless enamel adheres well. Plan your design on paper: remember that a lamp shade is round, so you probably want to make a "repeat design"—that is, a small pattern that appears around the shade several times. Transfer this design to the shade with pencil.

Illus. 47. The colors of heatless enamel you choose for your shade will glow when the light is turned on. The satiny smoothness cannot be equalled by any other medium.

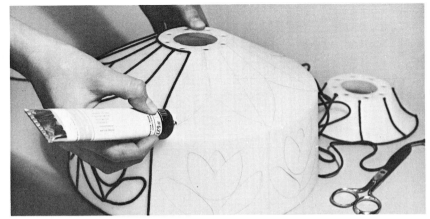

Divide the shade into sections both to separate each repeat pattern and to imitate stained glass by glueing black braid to the lamp. Use a clear-drying glue that does not corrode plastic. The braid not only separates the different colors to make the shade more attractive to look at, but it also acts as a border to prevent the wet colors from running into each other.

Illus. 49. Cut the braid to the proper length and place it on top of the glue. Press it with your fingers so it does not slip.

43

The colors may still run together, however, if you use liquid Polymer. No matter what the shape of your shade is, parts of the design are sure to fall in vertical areas, and liquid Polymer will drip off these sections. To prevent messy running, use heatless enamel GEL GLAZE in place of liquid Polymer. Mix Gel Glaze with Curing Agent in the proportion of 3 to 1.

When the glue holding the braid to the plastic is dry, apply the heatless enamel mixture to the shade with a stir stick. A brush is not stiff enough for the thick mixture. Push the enamel to the borders of each area, and try to make the surface even.

After you have filled in the colored sections, mix a sufficient quantity of Clear Extender with

Illus. 50. Gel Glaze, which is thicker than Polymer, does not pour very readily. Pour Curing Agent into the Glaze container.

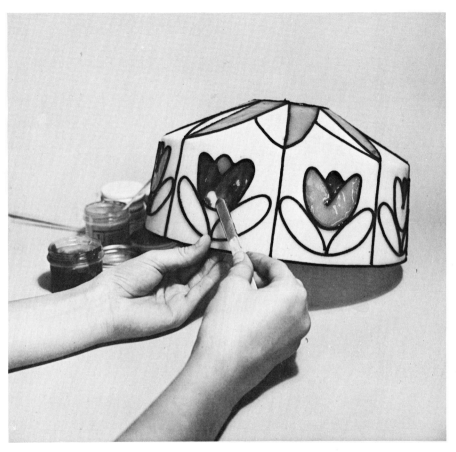

Illus. 51. Use a stir stick like a spatula to apply the Glaze/Curing Agent mixture to the side of the shade.

Illus. 52. For the top of the shade, and for large areas, it is easier to use liquid Polymer rather than Gel Glaze. Apply this mixture with a brush.

Curing Agent to brush over the rest of the shade for a background glaze. Use either a stir stick or a paint brush to apply this, bringing the mixture right up to the braid borders. Set the lamp shade in a warm dry room for about a week to harden completely. Once the heatless enamel has dried, you can use your shade over the hottest lamp— without worrying about melting, peeling or cracking.

Heatless enamel is remarkable for its long-lasting qualities on any surface, under any temperature. The possibilities of decoration that are open to you are limitless when you use this innovative product!

Illus. 53. Nature, holidays, food, sports—just look around you for ideas! Soon you will be adding attractive motifs to every surface in your home.

Index